i

VANISHING SUN: Behind the Amazing Science and Beauty of the Solar Eclipse

© 2023 by Dr. Leo Lexicon

VANISHING SUN
Behind the Amazing Science and Beauty of the Solar Eclipse

by

Dr. Leo Lexicon

VANISHING SUN
Behind the Amazing Science and Beauty of the Solar Eclipse

Are you ready to be dazzled?
"Vanishing Sun: Behind the Amazing Science of the Solar Eclipse"
is your ultimate guide to understanding and experiencing one of
nature's most breathtaking spectacles.

Imagine a world where the sun vanishes in the middle of the day,
and the sky darkens to an eerie twilight. It's not a scene from a
science fiction movie; it's a real celestial event that has captivated
humans for millennia. Solar eclipses are not just awe-inspiring;
they're also critical windows into the workings of our universe.

In this book, you will discover the fascinating science behind solar
eclipses, from the cosmic dance of the sun, moon, and Earth to the
cutting-edge research that scientists conduct during these rare
events. You'll learn about the myths and legends that have sprung
up around eclipses throughout history and the real-world impacts
they've had on human events.

With detailed guides on eye safety, eclipse photography, and the
best viewing locations for upcoming eclipses, you will be fully
prepared to witness this awe-inspiring phenomenon for yourself.
As we look to the future, solar eclipses offer more than just a
breathtaking spectacle; they're a reminder of the endless wonders
that await us in the sky. This book will inspire you to look up,
explore, and appreciate the marvels of our universe like never
before.

Dr. Leo Lexicon is an educator and author. He is the founder of
Lexicon Labs, a publishing imprint that is focused on creating
entertaining and educational books for active minds.

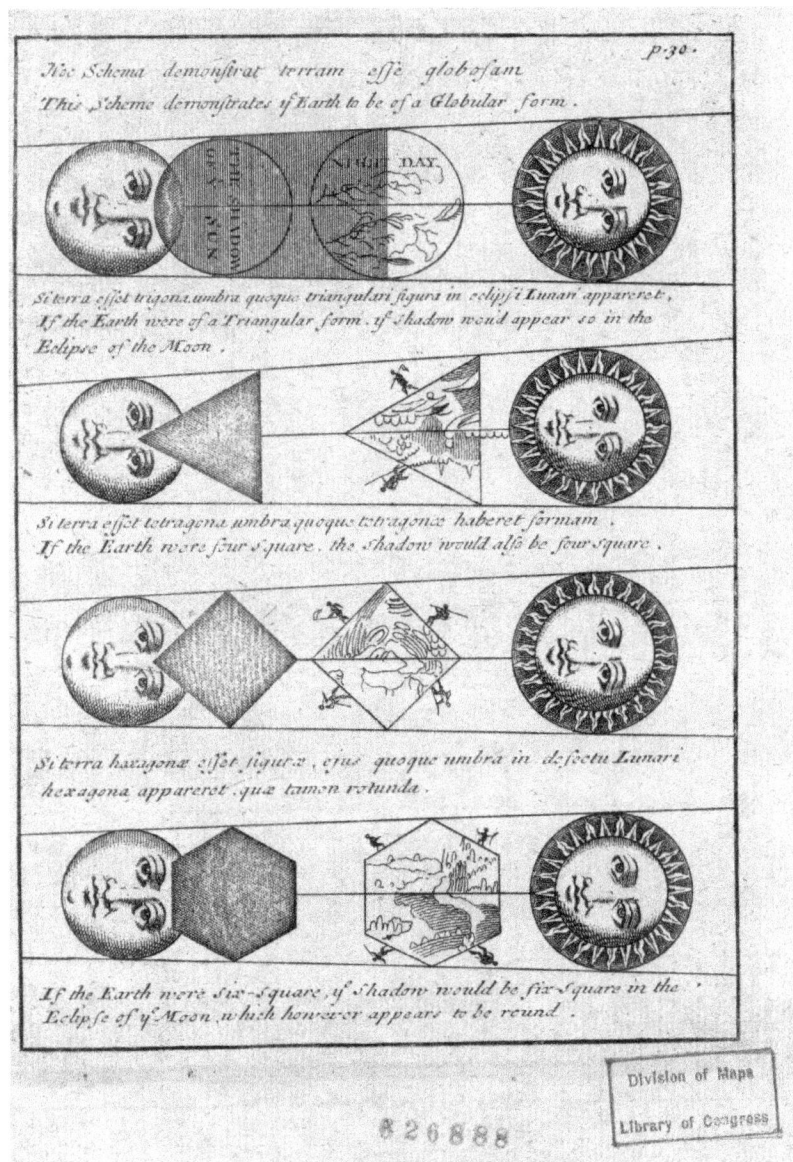

Four Diagrams of Solar Eclipses, Johanne Bunone, 1711
(Source: Library of Congress)

CONTENTS

Chapter 1: What is a Solar Eclipse?

When the Moon Blocks the Sun

Have you ever wondered what would happen if the sun suddenly disappeared from the sky? Imagine you're outside on a bright, sunny day when suddenly, the sky starts to darken. The birds stop singing, and the air gets cooler. The sunlight fades away until it's almost like nighttime in the middle of the day! This incredible event is called a solar eclipse, and it happens when the moon moves in front of the sun, blocking its light from reaching Earth. A solar eclipse is a rare and spectacular event that has fascinated people for thousands of years. It occurs when the moon passes between the sun and Earth, casting its shadow onto our planet's surface. During a total solar eclipse, the moon completely covers the sun's bright disk, allowing us to see its beautiful outer atmosphere called the corona. This only happens when the moon is at just the right distance from Earth and when it crosses paths with the sun at precisely the right time.

Let us now look at the different types of solar eclipses.

Total, Partial, and Annular Eclipses

There are three main types of solar eclipses: total, partial, and annular. The type of eclipse we see depends on where we are located within the moon's shadow and how perfectly aligned the sun, moon, and Earth are during the event.

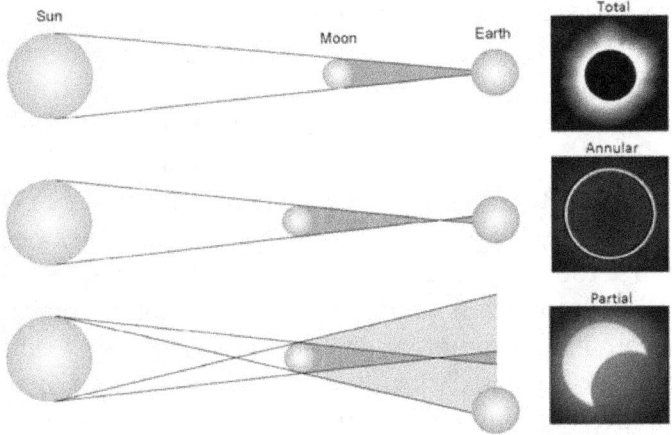

Fig. Types of Solar Eclipses

A total solar eclipse happens when the moon completely blocks out the sun's light from reaching a specific area on Earth's surface. This area is called the path of totality - a narrow strip of land usually about 100-160 kilometers (60-100 miles) wide - where people can see the total eclipse phase when day turns into an eerie twilight for a few minutes. During totality, it becomes dark enough for bright stars and planets to become visible in what was just moments ago a daytime sky! Total eclipses also reveal the sun's outer atmosphere called the corona - beautiful wisps of pearly white light extending far out into space around our star which remains hidden behind our atmosphere except during these special events.

Partial eclipses occur when only part of sun covered by moon so it looks kind of like someone took a bite out of side or top our star depending on your location relative moon's shadow path across earth at time you view this type eclipse happening overhead where ever your standing then as it progresses through different phases over next hour or two before returning back full sunlight again later in day once moon moved away from sun's position again allowing its rays to fully shine down once more heating up ground below making earth feel warm and comfortable again just like it was prior eclipse beginning earlier.

Annular eclipses are unique because they occur when moon appears smaller than sun as seen from earth due its distance away from us compared with sun's distance away from us too at certain times throughout year based on where both celestial bodies located along their respective orbital paths around each other and earth as well. So even though moon passes directly in front of sun like during total eclipse, it does not fully cover up sun's disk because its apparent size too small cause complete blackout effect like we see happening with total version this phenomenon taking place instead leaving behind a thin ring (or "annulus") of very bright sunlight still shining around the dark silhouette of moon at its center creating a mesmerizing "ring of fire" effect across sky above for all those lucky enough to witness this version eclipse taking place whenever it does decide to occur next somewhere in world depending on time of year and positioning of all three objects involved - sun, moon and earth - relative to one another at that particular moment in space and time.

Why Don't We Have Eclipses Every Month?

If the moon orbits Earth every month, then why don't we see a solar eclipse during every new moon phase when it passes between us and the sun? The answer lies in the fact that the moon's orbit is tilted about 5 degrees relative to Earth's orbit around the sun (which astronomers call the ecliptic). This tilt means that most of the time, the moon passes above or below the sun as seen from Earth, and its shadow misses our planet completely.

Solar eclipses can only happen during a new moon phase when the moon is directly between Earth and the sun. But for an eclipse to occur, the moon must also be crossing the ecliptic (Earth's orbital plane) at just the right time and place. This alignment is relatively rare because of the moon's tilted orbit, which is why we don't see solar eclipses every month.

In fact, solar eclipses happen somewhere on Earth about every 18 months on average. However, because the moon's shadow is relatively small compared to Earth's size, the path of totality only falls on a specific part of the planet's surface each time, and the same location may not experience a total solar eclipse again for hundreds of years!

The path of totality for total solar eclipses usually sweeps across Earth's surface in a narrow band, traveling from west to east due to our planet's rotation. This means that the total eclipse phase is only visible from a small portion of the planet each time it occurs. To see a total solar eclipse, you need to be in just the right place at just the right time - a fact that has inspired many eclipse chasers to travel the world in pursuit of these awe-inspiring celestial events. Partial eclipses, on the other hand, can be seen from a much larger area on either side of the path of totality, sometimes even from entire continents or hemisphere depending on the specific geometry of each eclipse. So while you may not always be able to see a total eclipse from your location, chances are you will have many opportunities in your lifetime to witness at least a partial solar eclipse if you keep an eye on the sky and know when and where to look.

In the next chapter, we will explore the exciting details surrounding the upcoming total solar eclipse of 2024 - including where and when you can see it, how long totality will last, and what makes this particular eclipse so special compared to others in recent history. So get ready to learn more about one of the most incredible astronomical events you can ever witness!

Chapter 2: The 2024 Solar Eclipse

The Path of Totality

Get ready for an astronomical adventure like no other! On April 8, 2024, a total solar eclipse will sweep across North America, plunging parts of Mexico, the United States, and Canada into an eerie daytime darkness. This extraordinary event will be visible along a narrow path called the "path of totality," which stretches from the Pacific to the Atlantic Ocean.

The path of totality is where the moon will completely cover the sun's bright disk, casting a dark shadow on Earth's surface. This shadow is called the umbra, and it will race across the continent at incredible speeds, covering a distance of over 3,000 miles (4,800 kilometers) in just 139 minutes! The width of the path will vary between 100 and 124 miles (160 and 200 kilometers), and only those lucky enough to be within this narrow strip will experience the full beauty and wonder of totality.

The eclipse will first make landfall on the Pacific coast of Mexico, near the city of Mazatlán, at around 11:07 a.m. local time. From there, it will quickly move northeastward, crossing into Texas near the city of Eagle Pass at around 1:27 p.m. CDT. The path of totality will then continue its journey across the United States, passing through parts of Oklahoma, Arkansas, Missouri, Illinois, Kentucky, Indiana, Ohio, Pennsylvania, New York, Vermont, New Hampshire, and Maine. In Canada, the eclipse will be visible in parts of Ontario, Quebec, New Brunswick, Prince Edward Island, and Newfoundland and Labrador before finally leaving the continent at around 4:18 p.m. NDT.

Where and When to See It

If you want to experience the total solar eclipse of 2024, you will need to plan ahead and choose your viewing location carefully. Some of the major cities that will lie within the path of totality include Mazatlán, Mexico; Austin and Dallas, Texas; Little Rock, Arkansas; Indianapolis, Indiana; Cleveland, Ohio; Buffalo and Rochester, New York; and Montreal, Quebec.

Keep in mind that weather conditions can greatly impact your eclipse viewing experience. Clear skies are essential for observing the eclipse, so it's a good idea to choose a location with a high probability of favorable weather. In the months leading up to the eclipse, astronomers and meteorologists will be closely monitoring weather patterns along the path of totality to help eclipse chasers make informed decisions about where to go.

Timing is also crucial when it comes to witnessing the total solar eclipse. The exact duration of totality will vary depending on your location within the path, with the longest period of darkness occurring near the center of the path. On average, totality will last around 4 minutes, but it can range from just a few seconds at the edges of the path to a maximum of 4 minutes and 28 seconds near Nazas, Durango, Mexico.

**Fig. A photo of the 2017 Total Solar Eclipse
(Source: NASA)**

To make the most of your eclipse experience, it's essential to know the precise times when the partial and total phases of the eclipse will occur at your chosen location. You can find detailed eclipse maps and timing information on websites such as NASA's Eclipse Website (eclipse.gsfc.nasa.gov) and the Great American Eclipse (greatamericaneclipse.com).

How Long Will It Last?

One of the most spectacular aspects of a total solar eclipse is the sudden onset of darkness during the daytime. As the moon gradually covers more and more of the sun's disk, the amount of sunlight reaching Earth's surface diminishes, and the sky begins to darken. In the minutes before totality, the landscape is bathed in an eerie, silvery light, and the remaining sliver of the sun takes on a crescent shape.

Just before the moon completely covers the sun, a dazzling display known as "Baily's beads" can be seen along the edge of the moon's silhouette. These beads of light are actually sunlight shining through the valleys and crevices along the moon's rugged surface. As the last bead disappears, a brilliant flash of light called the "diamond ring effect" marks the beginning of totality.

During totality, the sky darkens to a deep twilight, and stars and planets become visible in the daytime sky. The sun's faint outer atmosphere, called the corona, appears as a ghostly halo around the moon's dark disk. The corona is normally hidden by the sun's overwhelming brightness, but during a total eclipse, it can be seen stretching out into space, sometimes extending several times the diameter of the sun itself.

The duration of totality depends on several factors, including your location within the path of totality and the relative sizes and distances of the sun and moon at the time of the eclipse. For the 2024 eclipse, the maximum duration of totality will be 4 minutes and 28 seconds, which will occur near the town of Nazas, Durango, Mexico. In the United States, the longest duration of totality will be around 4 minutes and 26 seconds, visible from parts of Texas and Oklahoma.

As totality ends, the process reverses itself, with Baily's beads and the diamond ring effect appearing once again as the moon moves off the sun's disk. The sky gradually brightens, and the landscape returns to its normal daytime appearance. The entire eclipse event, from the beginning of the partial phase to the end, can last several hours, but the total phase itself is fleeting, making it all the more precious and awe-inspiring.

In the next chapter, we will dive into the fascinating science behind solar eclipses, exploring the celestial mechanics that make these events possible and the amazing phenomena that occur when the moon's shadow falls upon Earth. Get ready to learn about the intricate dance of shadows and light that creates one of nature's most breathtaking spectacles!

DR. LEO LEXICON

Chapter 3: The Science of Shadows

Relative Sizes of the Moon, Earth and Sun

Before we get into the details of an eclipse, let us first understand the relative sizes of the entities we are talking about. The sizes of the Moon, Earth, and Sun can be compared using their diameters as a common measure. Here are the average diameters of each celestial body:

- **Moon:** The Moon's average diameter is about 3,474 kilometers (2,159 miles).
- **Earth:** The Earth's average diameter is about 12,742 kilometers (7,918 miles).
- **Sun:** The Sun's average diameter is approximately 1,391,000 kilometers (864,000 miles).

To express the size ratio of the Moon, Earth, and Sun, we can simplify their relationship by dividing their diameters:

- **Moon to Earth Ratio:** To find the ratio of the Moon's size to the Earth's, we divide the Moon's diameter by the Earth's diameter.
- **Earth to Sun Ratio:** Similarly, to find the ratio of the Earth's size to the Sun's, we divide the Earth's diameter by the Sun's diameter.

Let us calculate these ratios for a clearer understanding. We find that the ratio of the sizes (diameters) of the Moon to the Earth is approximately **0.273**, and the ratio of the sizes of the Earth to the Sun is approximately **0.0092**.

This means that the Moon's diameter is about 27.3% that of the Earth, and the Earth's diameter is about 0.92% that of the Sun. These ratios highlight the vast differences in scale between these celestial bodies, with the Sun being significantly larger than both the Earth and the Moon.

How Shadows Work

To understand the science behind a solar eclipse, we must first explore the fascinating world of shadows. Shadows are a part of our everyday lives, but have you ever stopped to think about how they work? A shadow is created when an object blocks light from a source, preventing it from reaching a surface behind the object. The size, shape, and darkness of a shadow depend on several factors, including the size of the light source, the size and shape of the object blocking the light, and the distance between the light source, the object, and the surface where the shadow is cast.
In the case of a solar eclipse, the moon is the object that blocks sunlight from reaching Earth's surface. The sun is our light source, and it is much much larger than the moon. In fact, the sun's diameter is about 400 times larger than the moon's, making the moon a tiny speck in the vastness of our solar system! Also, when compared to the sun, the earth itself is very small. However, the moon is also about 400 times closer to Earth than the sun, which means that from our perspective, the sun and moon appear to be almost the same size in the sky. A hard concept to understand when we look at the picture below!

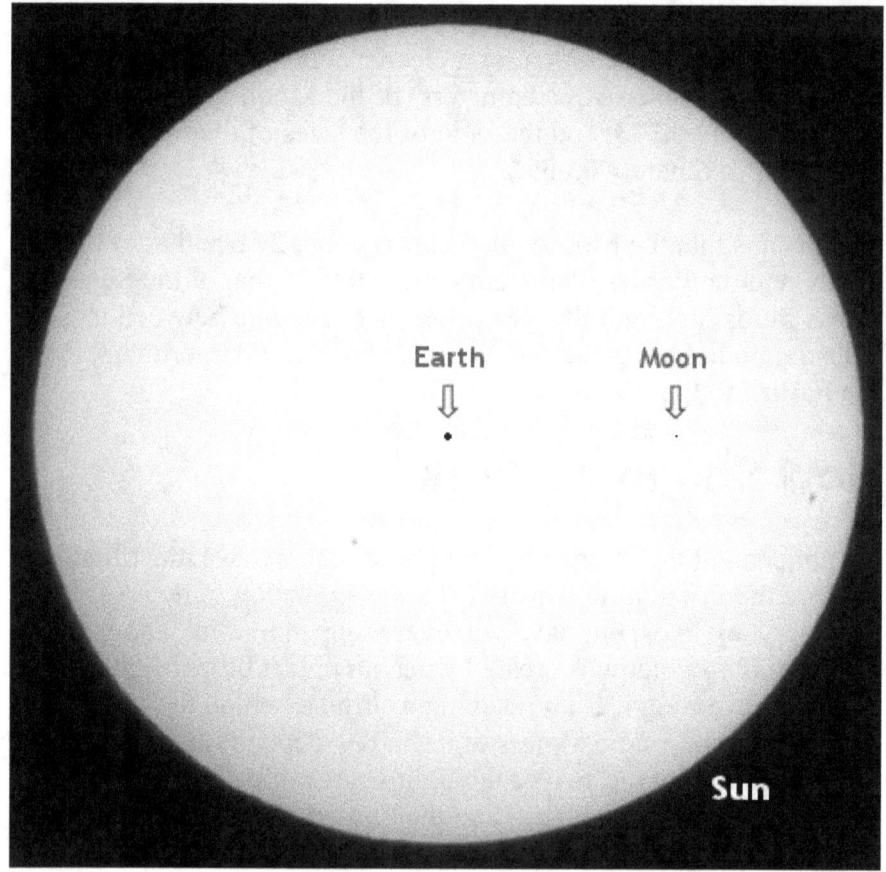

Fig. The Sun, Earth and Moon to Scale
(Source: Tom Roelandts)

This cosmic coincidence is what makes solar eclipses possible. When the moon passes directly between Earth and the sun, it can block out most or all of the sun's light, casting a shadow on Earth's surface. The type of shadow we see depends on where we are located relative to the moon's shadow and how well-aligned the sun, moon, and Earth are during the eclipse.

Umbra, Penumbra, and Antumbra

There are three distinct parts to the moon's shadow during a solar eclipse: the umbra, penumbra, and antumbra. Each of these shadow regions has unique characteristics that determine the type of eclipse visible from Earth's surface.

The umbra is the darkest part of the moon's shadow, where the moon completely blocks the sun's light. If you're standing within the path of the umbra, you will experience a total solar eclipse, with the moon perfectly covering the sun's disk and the sun's faint outer atmosphere, the corona, visible around the moon's silhouette. The penumbra is the much larger, lighter part of the moon's shadow that surrounds the umbra. Within the penumbra, the moon only partially blocks the sun's light, resulting in a partial solar eclipse. If you're located in the penumbra, you will see the moon taking a "bite" out of the sun's disk, but you won't experience the total darkness and the corona's appearance that those in the umbra will witness.

The antumbra is a unique part of the moon's shadow that comes into play during annular solar eclipses. An annular eclipse occurs when the moon is at its farthest point from Earth in its orbit, making it appear slightly smaller than the sun in the sky. As a result, the moon doesn't completely cover the sun's disk during an annular eclipse, leaving a bright ring of sunlight visible around the moon's silhouette. The antumbra is the region where this "ring of fire" effect is visible, and it extends beyond the umbra's tip.

Measuring the Size of the Moon's Shadow

The size of the moon's shadow on Earth's surface depends on several factors, including the relative sizes and distances of the sun and moon, as well as Earth's curvature. During a total solar eclipse, the umbra's diameter on Earth's surface can range from about 100 to 160 kilometers (60 to 100 miles), while the penumbra can be several thousand kilometers wide.

The length of the moon's shadow is also impressive, stretching out into space for nearly 1.4 million kilometers (870,000 miles)! However, because of Earth's curvature, only a small portion of this shadow actually falls on our planet's surface during an eclipse. Scientists use a variety of methods to measure the size and shape of the moon's shadow during a solar eclipse. One common technique involves using a network of ground-based telescopes and cameras to capture images of the moon's shadow as it moves across Earth's surface. These images can be combined to create detailed maps of the umbra and penumbra, allowing researchers to study the eclipse's characteristics and refine their predictions for future events.

Satellite imagery is another valuable tool for measuring the moon's shadow during a solar eclipse. NASA's Earth-observing satellites, such as Terra and Aqua, have captured stunning images of the moon's shadow sweeping across Earth's surface during past eclipses. These images not only provide a unique perspective on the eclipse itself but also help scientists study how the moon's shadow affects Earth's atmosphere and weather patterns. In addition to ground-based and satellite observations, scientists also use computer models to simulate solar eclipses and predict the size, shape, and path of the moon's shadow. These models take into account a wide range of factors, including the precise positions and motions of the sun, moon, and Earth, as well as the effects of Earth's atmosphere and surface features on the shadow's appearance.

By combining observational data with theoretical models, scientists can create incredibly accurate maps of the moon's shadow and predict the timing and location of future solar eclipses with remarkable precision. This information is essential for eclipse chasers, as it helps them plan their expeditions and ensures they are in the right place at the right time to witness these awe-inspiring celestial events.

In the next chapter, we will explore the intricate orbital dance of the sun, moon, and Earth that makes solar eclipses possible, and learn how the tilt of Earth's axis and the cycles of the moon's orbit determine when and where these events occur. Get ready to dive into the fascinating world of celestial mechanics and discover the secrets behind one of nature's most captivating phenomena!

Chapter 4: The Sun, Moon and Earth Dance

The Orbits of the Earth and Moon

To understand how solar eclipses occur, we need to take a closer look at the intricate dance between the sun, moon, and Earth. This cosmic ballet is governed by the laws of gravity and the orbital paths each body follows as they move through space.

Let's start with Earth's orbit around the sun. Our planet follows an elliptical path, taking approximately 365.25 days to complete one full revolution. This is why we have leap years every four years, to account for that extra quarter of a day. Earth's orbit is not a perfect circle, but it's very close, with the distance between the sun and Earth varying by only about 3% throughout the year.

As Earth orbits the sun, the moon orbits Earth. The moon's orbit is also elliptical, with an average distance of about 384,400 kilometers (238,900 miles) from Earth. It takes the moon approximately 27.3 days to complete one orbit around Earth with respect to the background stars. However, because Earth is also moving around the sun during this time, the moon needs a little extra time to catch up and complete a full cycle of phases, which takes about 29.5 days.

The moon's orbit is tilted by about 5 degrees relative to Earth's orbit around the sun. This tilt is crucial for understanding why we don't have solar eclipses every month. If the moon's orbit were perfectly aligned with Earth's orbital plane, we would indeed experience a solar eclipse every new moon and a lunar eclipse every full moon. However, due to the tilt, the moon usually passes above or below the sun from Earth's perspective, and its shadow misses our planet entirely.

How the Tilt of Earth's Axis Affects Eclipses

Earth's axis of rotation is tilted by about 23.5 degrees relative to its orbital plane around the sun. This tilt is what gives us our seasons, as different parts of the planet receive more or less direct sunlight throughout the year. But how does this tilt affect solar eclipses?

The tilt of Earth's axis, combined with the tilt of the moon's orbit, creates a unique set of circumstances that determine when and where solar eclipses can occur. For a solar eclipse to happen, the moon must be in the new moon phase and aligned with the plane of Earth's orbit around the sun (known as the ecliptic). This alignment occurs at two points in the moon's orbit, called nodes.

The nodes are the points where the moon's orbital plane crosses the ecliptic. When the moon is near a node during the new moon phase, its shadow can fall on Earth's surface, resulting in a solar eclipse. The type of eclipse (total, partial, or annular) depends on how closely aligned the sun, moon, and Earth are at that moment.

The tilt of Earth's axis also affects the timing and location of solar eclipses. During the Northern Hemisphere's summer months, the North Pole is tilted towards the sun, placing the Northern Hemisphere closer to the sun than the Southern Hemisphere. This means that solar eclipses occurring during this time are more likely to be visible from the Northern Hemisphere. Conversely, during the Southern Hemisphere's summer months, the South Pole is tilted towards the sun, making solar eclipses more likely to be visible from the Southern Hemisphere.

The Saros Cycle: When Eclipses Repeat

One of the most fascinating aspects of solar eclipses is their predictability. Astronomers can calculate the dates and locations of future eclipses with incredible accuracy, thanks to a remarkable pattern known as the Saros cycle.

The Saros cycle is a period of approximately 18 years, 11 days, and 8 hours, during which a specific sequence of eclipses repeats itself. This means that if you witness a particular solar eclipse, a very similar eclipse will occur one Saros period later, with the path of totality shifted about 120 degrees westward around Earth's surface.

The Saros cycle arises from a near-resonance between three key orbital periods: the synodic month (the time it takes for the moon to complete a full cycle of phases, about 29.5 days), the draconic month (the time it takes for the moon to return to the same node in its orbit, about 27.2 days), and the anomalistic month (the time it takes for the moon to return to its closest point to Earth, or perigee, about 27.6 days).

When these three periods align, the sun, moon, and Earth return to nearly the same relative positions they held at the beginning of the cycle. This alignment results in a repeating pattern of eclipses, with each eclipse in the sequence separated by one Saros period.

A Saros series begins with a partial eclipse near one of Earth's polar regions, with subsequent eclipses becoming more central and moving towards the equator over time. The series typically lasts between 1,226 and 1,550 years, producing 69 to 87 eclipses before ending with a partial eclipse near the opposite polar region.

Understanding the Saros cycle is crucial for predicting future eclipses and planning expeditions to witness these events. By knowing the date and location of a previous eclipse in a Saros series, astronomers can calculate when and where the next eclipse in the series will occur, helping eclipse chasers position themselves in the path of totality.

The Saros cycle also helps scientists study long-term changes in Earth's rotation and orbit. By comparing the timing and location of ancient eclipse records with modern calculations, researchers can detect minute changes in Earth's spin rate and orbit over centuries and millennia.

In the next chapter, we will explore the fascinating history of eclipse observations and the valuable scientific insights gained from studying these celestial events. From ancient astronomers to modern-day researchers, the pursuit of understanding solar eclipses has driven remarkable advancements in our knowledge of the universe and our place within it. Get ready to embark on a journey through time and discover how eclipse chasers and scientists have unraveled the mysteries of the sun, moon, and Earth's cosmic dance!

Chapter 5: Eclipse Chasers and Scientists

Famous Eclipse Expeditions in History

Throughout history, solar eclipses have captivated the minds of people across the globe. From ancient civilizations to modern-day scientists, the allure of these celestial events has inspired countless expeditions, scientific discoveries, and cultural traditions. In this chapter, we will explore some of the most famous eclipse expeditions in history and the valuable insights they have provided into the workings of our universe.

One of the earliest recorded eclipse expeditions dates back to 585 BCE, when the Greek philosopher Thales of Miletus allegedly predicted a solar eclipse that ended a battle between the Lydians and the Medes in ancient Anatolia (modern-day Turkey). According to the ancient historian Herodotus, the warring parties were so amazed by the sudden darkness that they ceased fighting and negotiated a peace treaty. While the accuracy of this account is debated, it highlights the profound impact solar eclipses have had on human history.

Fast forward to the 19th century, and we find one of the most famous eclipse expeditions of all time. In 1919, British astronomer Sir Arthur Eddington led an expedition to the island of Príncipe off the west coast of Africa to observe a total solar eclipse. Eddington's goal was to test Albert Einstein's theory of general relativity, which predicted that massive objects like the sun would warp the fabric of space-time, bending the path of light from distant stars.

During the eclipse, Eddington photographed the positions of stars near the sun's edge and compared them to photographs taken of the same stars at night, when the sun was not present. The results showed that the stars' positions had indeed shifted, confirming Einstein's predictions and providing the first observational evidence for general relativity. This groundbreaking discovery revolutionized our understanding of gravity and the nature of space and time.

In more recent years, eclipse expeditions have continued to yield valuable scientific insights. In 1973, a team of scientists led by American astronomer John Eddy used a solar eclipse to study the sun's outermost atmosphere, known as the corona. By analyzing the corona's spectrum during totality, Eddy and his colleagues discovered that the corona was much hotter than previously believed, reaching temperatures of several million degrees Celsius. This finding challenged existing theories about the sun's energy transfer mechanisms and sparked new research into the complex processes driving the corona's extreme heat.

What Scientists Learn from Eclipses

Solar eclipses provide unique opportunities for scientists to study the sun and its effects on Earth's atmosphere and environment. During totality, the moon blocks out the sun's bright disk, allowing researchers to observe the sun's faint outer atmosphere, the corona, which is normally obscured by the sun's overwhelming glare. By studying the corona during eclipses, scientists can gain valuable insights into the sun's magnetic field, solar wind, and the processes that heat the corona to extremely high temperatures. These observations help us better understand the sun's influence on Earth, from the formation of auroras to the potential impact of solar storms on our technology and infrastructure.

Eclipses also allow scientists to study Earth's upper atmosphere and ionosphere, the charged layer of particles that plays a crucial role in radio communication and satellite navigation. During an eclipse, the sudden absence of sunlight causes rapid changes in the ionosphere's composition and density, affecting the propagation of radio waves. By monitoring these changes, researchers can improve our understanding of the ionosphere's dynamics and develop better models for predicting its behavior.

Additionally, solar eclipses provide opportunities to study the response of Earth's ecosystems to sudden changes in light and temperature. Scientists have observed how animals and plants react to the onset of darkness during totality, with some species exhibiting unusual behaviors or changes in activity patterns. These observations can provide valuable insights into the adaptability and resilience of different organisms to environmental changes.

Citizen Science: How You Can Contribute

While professional scientists lead many eclipse expeditions and research projects, amateur astronomers and citizen scientists can also make valuable contributions to our understanding of solar eclipses. Citizen science projects allow individuals from all walks of life to participate in scientific research, collecting data and making observations that can help answer important questions about the sun, moon, and Earth.

One example of a citizen science project focused on solar eclipses is the Eclipse Megamovie Project, which aimed to create a high-resolution video of the 2017 total solar eclipse across the United States. The project enlisted the help of thousands of volunteers, who captured images of the eclipse from different locations along the path of totality. By stitching together these images, scientists created a continuous video of the eclipse's progress, revealing intricate details of the sun's corona and the moon's surface.

Another way citizen scientists can contribute to eclipse research is through the observation and recording of solar phenomena, such as Bailey's beads, shadow bands, and the diamond ring effect. These phenomena occur in the moments just before and after totality and can provide valuable information about the sun's structure and the moon's topography. By carefully documenting these observations and sharing them with the scientific community, amateur astronomers can help refine our models of solar eclipses and improve our predictions for future events.

Citizen scientists can also assist in monitoring the effects of solar eclipses on Earth's environment, such as changes in temperature, wind patterns, and animal behavior. By setting up weather stations, recording wildlife activity, or even simply noting the changes in light and shadow during an eclipse, individuals can contribute to our understanding of how these events impact our planet's ecosystems.

Participating in citizen science projects not only helps advance scientific knowledge but also fosters a sense of connection and engagement with the natural world. By taking part in eclipse observations and research, individuals can gain a deeper appreciation for the beauty and complexity of our universe, while also contributing to the collective effort to understand its workings.

As we look ahead to the 2024 total solar eclipse and beyond, there will be numerous opportunities for citizen scientists to get involved and make their mark on eclipse research. Whether through organized projects or individual observations, the contributions of amateur astronomers and enthusiasts will continue to play a vital role in unraveling the mysteries of solar eclipses and their impact on our world.

In the next chapter, we will explore the practical aspects of preparing for an eclipse, including eye safety, equipment recommendations, and tips for planning your eclipse viewing experience. By arming yourself with the right knowledge and tools, you will be ready to witness one of nature's most awe-inspiring spectacles and potentially contribute to the scientific understanding of these celestial events. Get ready to embark on your own eclipse adventure!

Citizen Science Resources

The initiatives mentioned below allow the public to collaborate with researchers, making valuable contributions to our understanding of this celestial phenomenon.

- **Eclipse Megamovie 2024**
  (https://eclipsemegamovie.org/): This NASA-funded project is gearing up for the 2024 total solar eclipse. It invites citizen scientists to capture high-resolution images of the eclipse using cameras on special mounts. These images will be combined to create a stunning mosaic, revealing the sun's corona in unprecedented detail.

- **Eclipse Soundscapes Project**
  (https://eclipsesoundscapes.org/): This project, funded by NASA Science Activation, explores how wildlife reacts to solar eclipses. You can participate from anywhere in the world! The project is seeking observations of animal behavior during the April 2024 eclipse, and sound recordings can be submitted by those lucky enough to be in the path of totality.

Beyond these specific projects, here are some additional resources to connect with the solar eclipse citizen science community:
- **Facebook Groups**: Search for groups dedicated to astronomy, citizen science, or eclipses. These groups often share information about ongoing projects and provide a platform to connect with other enthusiasts.

- **Reddit Threads**: Subreddits like r/Astronomy and r/CitizenScience can be great resources for finding discussions and updates on eclipse-related citizen science projects.

- **SciStarter (https://scistarter.org/):** This website is a comprehensive directory of citizen science projects across various disciplines. You can search for projects based on your interests and location, including those related to solar eclipses.

By participating in these citizen science projects, you can play an active role in scientific discovery and gain a deeper appreciation for the awe-inspiring phenomenon of a solar eclipse.

Chapter 6: Prepare for Eclipse Day

Eye Safety: Protecting Your Vision

As you prepare to witness the breathtaking spectacle of a solar eclipse, it's crucial to prioritize eye safety. Looking directly at the sun, even during a partial eclipse, can cause serious and permanent damage to your eyes. The sun's intense ultraviolet (UV) and infrared (IR) radiation can harm the retina, leading to a condition called solar retinopathy, which can result in temporary or lasting vision loss.

To safely observe a solar eclipse, you must use proper eye protection designed specifically for viewing the sun. Regular sunglasses, even those with very dark lenses, are not sufficient to protect your eyes during an eclipse. Instead, you should use certified solar eclipse glasses or handheld solar viewers that meet the ISO 12312-2 international safety standard.

These special glasses are made with filters that block out 100% of the sun's harmful UV and IR rays, as well as 99.99% of its intense visible light. They allow you to view the sun safely during all phases of a solar eclipse, except during totality when the moon completely covers the sun's disk.

When selecting solar eclipse glasses, be sure to purchase them from reputable vendors and check for the ISO 12312-2 certification printed on the glasses or packaging. Avoid using glasses that are more than three years old, as the protective filters can degrade over time. Also, inspect your glasses before use, and discard them if you notice any scratches, punctures, or damage to the lenses.

During the partial phases of the eclipse, when any part of the sun's disk is visible, you must wear your solar eclipse glasses continuously while looking at the sun. Remove the glasses only during totality, when the moon completely blocks the sun's bright face. As soon as the first glimmer of sunlight reappears at the end of totality, known as the "diamond ring effect," immediately put your glasses back on to avoid eye damage.

In addition to solar eclipse glasses, you can use other methods to safely view the eclipse, such as pinhole projectors or solar filters attached to telescopes or binoculars. These methods allow you to observe the eclipse indirectly, projecting an image of the sun onto a screen or piece of paper.

Remember, the most important aspect of eclipse viewing is ensuring your eye safety. By following these guidelines and using proper protective equipment, you can enjoy the awe-inspiring beauty of a solar eclipse without risking your eyesight.

Making a Pinhole Projector

While solar eclipse glasses provide a direct and convenient way to view an eclipse, you can also safely observe this celestial event using a simple pinhole projector. This method allows you to project an image of the sun onto a screen, enabling you to see the eclipse's progress without looking directly at the sun.

To make a basic pinhole projector, you will need two pieces of stiff, white cardboard or thick paper, a pin or needle, and a sunny location. Cut a small, square hole in the center of one of the cardboard pieces, about 1 to 2 inches wide. Then, use the pin to poke a small hole in the center of the square opening.

Take your pinhole projector outside and stand with your back to the sun. Hold up the cardboard with the pinhole, allowing sunlight to pass through the hole and onto the second piece of cardboard, which will serve as your projection screen. Adjust the distance between the two cardboard pieces until you see a sharp image of the sun projected onto the screen.

As the eclipse progresses, you will see the moon gradually blocking out more and more of the sun's disk on the projected image. You can experiment with different distances between the pinhole and the screen to change the size of the projected image, but remember that the image will be inverted due to the way light passes through the pinhole.

You can also create more advanced pinhole projectors using a cardboard box, aluminum foil, and a white sheet of paper. Cut a rectangular hole in one end of the box and tape a piece of aluminum foil over the opening. Use a pin to create a small hole in the center of the foil. At the opposite end of the box, cut a viewing hole and tape a white sheet of paper inside the box to serve as your projection screen.

Aim the pinhole end of the box towards the sun, allowing sunlight to enter through the pinhole and project an image of the eclipse onto the paper screen inside the box. This enclosed design provides a darker viewing environment, making the projected image easier to see.

Pinhole projectors offer a safe and inexpensive way to observe a solar eclipse, and they can be a fun and educational activity for children and adults alike. By experimenting with different designs and materials, you can create your own unique eclipse viewing experience while keeping your eyes protected from the sun's harmful rays.

Planning Your Eclipse Viewing Experience

To make the most of your solar eclipse experience, it's essential to plan ahead and choose a suitable viewing location. Whether you're traveling to the path of totality or observing a partial eclipse from your hometown, taking the time to prepare will ensure a memorable and enjoyable experience.

First, research the eclipse path and timing for your location. Use reliable sources, such as NASA's interactive eclipse maps or the Eclipse Wise website, to determine the best places to view the eclipse and the exact times for each phase of the event. Keep in mind that weather conditions can greatly impact eclipse viewing, so consider locations with a higher probability of clear skies.

If you plan to travel to the path of totality, make your arrangements well in advance. Accommodations and transportation can fill up quickly in popular eclipse viewing areas, so book early to secure your spot. When selecting a viewing site, look for open spaces with unobstructed views of the sky, such as parks, fields, or hilltops.

In the days leading up to the eclipse, monitor weather forecasts for your chosen location. While you can't control the weather, having a backup plan or being prepared to relocate on short notice can help ensure you don't miss the eclipse due to cloudy skies.

On eclipse day, arrive at your viewing site early to set up your equipment and find a comfortable spot. Be sure to have your solar eclipse glasses or other safe viewing devices ready, as well as any cameras, binoculars, or telescopes you plan to use. If you're using a pinhole projector, take the time to test it out and make any necessary adjustments before the eclipse begins.

As the eclipse progresses, take note of the changes in your surroundings. During a partial eclipse, you may notice a slight drop in temperature and a shift in the quality of the light. If you're in the path of totality, be prepared for a more dramatic experience, with the sky darkening to a deep twilight, stars and planets becoming visible, and the sun's corona shimmering around the edges of the moon.

Remember to follow eye safety guidelines throughout the eclipse, using your solar eclipse glasses or other protective devices whenever any part of the sun's disk is visible. During totality, it is safe to remove your glasses and view the eclipse directly, but be sure to put them back on as soon as the first glimmer of sunlight reappears.

In addition to observing the eclipse itself, consider documenting your experience through photographs, videos, or sketches. These personal records can help you remember the event and share your excitement with others. However, don't spend the entire eclipse behind a camera – be sure to take time to fully immerse yourself in the moment and appreciate the beauty and wonder of this celestial event.

By planning ahead, choosing a suitable viewing location, and following safety guidelines, you will be well-prepared to enjoy the 2024 solar eclipse and create lasting memories of this once-in-a-lifetime experience.

Chapter 7: Eclipses in Myth and Memory

Ancient Explanations for Eclipses

Throughout human history, solar eclipses have been met with a mix of fear, fascination, and awe. In the absence of scientific understanding, ancient civilizations created myths and legends to explain these mysterious and seemingly supernatural events. Many ancient cultures believed that solar eclipses were caused by celestial beings or animals devouring or stealing the sun. In ancient China, for example, people believed that a heavenly dog or dragon was responsible for eating the sun during an eclipse. To frighten the creature away and restore the sun's light, they would bang on drums, pots, and pans, creating a cacophony of noise. Similarly, in Norse mythology, eclipses were thought to occur when two wolves, Sköll and Hati, chased and caught the sun or moon. The Vikings believed that the end of the world, or Ragnarök, would come when the wolves finally succeeded in devouring the celestial bodies.

In Hindu mythology, the demon Rahu, who had been decapitated by the god Vishnu, was said to seek revenge by swallowing the sun or moon, causing an eclipse. According to legend, Rahu's severed head would consume the celestial body, but because he had no body, the sun or moon would eventually reappear, slipping out from his throat.

The ancient Greeks also had several mythological explanations for solar eclipses. One tale involved the young Thessalian witch Aglaonice, who claimed to have the power to draw down the moon from the sky. Some scholars suggest that Aglaonice may have used her knowledge of eclipse cycles to trick people into believing in her supernatural abilities.

In Mesoamerican cultures, such as the Aztecs and Mayans, solar eclipses were often associated with the battles between celestial deities. The Aztecs believed that during a solar eclipse, the moon goddess Coyolxauhqui was attacking her brother, the sun god Huitzilopochtli. The Mayans, on the other hand, saw eclipses as a time of great danger, when the Jaguar God of the Underworld would swallow the sun.

While these ancient explanations for eclipses may seem fanciful or superstitious to modern audiences, they reflect the deep significance that these celestial events held for early civilizations. Eclipses were seen as powerful omens, capable of influencing the fate of nations and individuals alike.

Eclipse Legends from Around the World

In addition to the well-known myths from ancient civilizations, countless eclipse legends and folktales have emerged from cultures around the globe. These stories offer a fascinating glimpse into the diverse ways that people have interpreted and responded to solar eclipses throughout history.

In some African cultures, eclipses were seen as a time of reconciliation and forgiveness. The Batammaliba people of Benin and Togo believed that during an eclipse, old feuds should be forgotten, and enemies should come together to resolve their differences. They saw the darkness of the eclipse as a metaphor for the dark feelings that can cloud human relationships, and the return of the sun's light as a symbol of the restoration of harmony and understanding.

In parts of Australia, Aboriginal legends often portrayed solar eclipses as a time of transformation and renewal. One tale from the Euahlayi people tells of a sun goddess who periodically swallows the moon, causing an eclipse. During this time, the goddess is said to be reborn, shedding her old, wrinkled skin and emerging fresh and young once more.

Some Native American cultures saw eclipses as a sign of the sun's displeasure or a warning of impending disaster. The Pomo people of California believed that a solar eclipse signaled that the sun was angry and might withhold its light and warmth from the earth. To appease the sun, they would sing, dance, and offer prayers and sacrifices.

In Inuit folklore, eclipses were sometimes attributed to the actions of sky deities or celestial animals. One legend tells of a sky goddess named Malina, who fled from her cruel brother, the moon god Anningan. As Anningan chased Malina across the sky, he would occasionally catch up to her, causing a solar or lunar eclipse.

These diverse eclipse legends serve as a testament to the enduring power and mystery of these celestial events. While modern science has provided us with a clear understanding of the mechanics behind eclipses, the sense of wonder and awe that they inspire remains unchanged. By exploring these ancient stories and myths, we can gain a deeper appreciation for the ways in which eclipses have captured the human imagination throughout history.

How Eclipses Have Affected Human Events

In addition to their cultural and mythological significance, solar eclipses have also played a role in shaping the course of human history. From ancient battles to modern scientific expeditions, eclipses have sometimes had a surprising influence on the outcome of important events.

One famous example is the eclipse of May 28, 585 BCE, which is said to have ended a long-standing war between the Lydians and the Medes in ancient Anatolia (modern-day Turkey). According to the ancient Greek historian Herodotus, the two armies were in the midst of a fierce battle when the day suddenly turned to night, causing the soldiers to lay down their weapons in awe and terror. The warring kings, Alyattes of Lydia and Cyaxares of Media, took the eclipse as a sign from the gods to end their conflict and agreed to a truce.

While the accuracy of Herodotus' account is debated, the story highlights the profound psychological impact that a sudden and unexpected eclipse could have on people in the ancient world. In an era when celestial events were often seen as omens or messages from the gods, an eclipse could be interpreted as a powerful sign of divine intervention.

More recently, solar eclipses have played a role in advancing scientific knowledge and understanding. In 1919, a team of astronomers led by Sir Arthur Eddington used observations of a total solar eclipse to test Albert Einstein's theory of general relativity. By comparing the apparent positions of stars near the eclipsed sun to their usual positions in the night sky, Eddington was able to confirm that the sun's gravity did indeed bend the path of starlight, just as Einstein had predicted. This groundbreaking experiment helped to establish general relativity as one of the foundational theories of modern physics.

Solar eclipses have also been used to study the sun's outer atmosphere, or corona, which is normally obscured by the sun's bright disk. During a total eclipse, the moon blocks out the sun's main light, allowing scientists to observe the delicate structure and composition of the corona. These observations have led to important discoveries about the sun's magnetic field, solar wind, and the processes that heat the corona to extremely high temperatures.

In addition to their scientific significance, eclipses have sometimes had unexpected effects on human events. During the total solar eclipse of August 11, 1999, which was visible across much of Europe, several power grids experienced sudden drops in electrical demand as millions of people stepped outside to watch the eclipse. This sudden change in power consumption caused some electrical generators to shut down unexpectedly, leading to brief power outages in a few areas.

Similarly, during the total solar eclipse of March 29, 2006, which was visible across parts of Africa and Asia, the sudden drop in solar radiation caused a temporary cooling of the earth's atmosphere. This cooling effect, combined with changes in atmospheric pressure and wind patterns, may have contributed to a brief but intense thunderstorm that struck the city of Accra, Ghana, just minutes after the eclipse ended.

These examples demonstrate the complex and sometimes surprising ways in which solar eclipses can influence human events and the natural world. By studying these celestial phenomena and their impacts, we can gain a deeper understanding of the intricate web of connections between the sun, the moon, and life on earth.

Chapter 8: Other Amazing Phenomena

Solar and Lunar Halos

While solar eclipses are undoubtedly the most spectacular celestial events involving the sun and moon, there are several other fascinating phenomena that can be observed in the sky. One such phenomenon is the appearance of solar and lunar halos, which are caused by the refraction and reflection of light by ice crystals in the atmosphere.

Solar halos are typically seen as bright, circular rings around the sun, often with a radius of about 22 degrees. These halos form when sunlight passes through hexagonal ice crystals in high-altitude cirrus clouds. The ice crystals act like tiny prisms, bending the light and creating a halo effect. Sometimes, depending on the orientation of the ice crystals, additional arcs or spots of light, known as sun dogs or parhelia, can appear on either side of the sun.

Lunar halos are similar to solar halos but form around the moon instead. They are most commonly seen as a pale, whitish ring encircling the moon, with a radius of about 22 degrees. Like solar halos, lunar halos are caused by the refraction of moonlight through ice crystals in thin, high-altitude clouds.

In folklore, the appearance of a lunar halo was often interpreted as a sign of impending rain or snow. While halos themselves do not directly cause precipitation, they can indicate the presence of high-altitude moisture, which may eventually lead to rain or snow if other atmospheric conditions are favorable.

Both solar and lunar halos are relatively common phenomena, and unlike solar eclipses, they can be observed from many locations around the world. However, the brightness and visibility of halos can vary depending on factors such as the position of the sun or moon, the density and structure of the ice crystals, and the overall clarity of the sky.

Supermoons and Micromoons

Another fascinating phenomenon related to the moon is the occurrence of supermoons and micromoons. These terms refer to the apparent size of the moon as seen from Earth, which can vary slightly due to the moon's elliptical orbit.

A supermoon occurs when the moon is at its closest point to Earth (perigee) and coincides with a full moon phase. During a supermoon, the moon appears larger and brighter than usual, as it is about 7% closer to Earth than the average distance. Supermoons tend to be more noticeable when the moon is near the horizon, due to an optical illusion known as the "moon illusion," which makes the moon appear even larger in comparison to nearby objects.

Conversely, a micromoon happens when the moon is at its farthest point from Earth (apogee) and coincides with a full moon phase. During a micromoon, the moon appears slightly smaller and less bright than usual, as it is about 7% farther away from Earth than the average distance.

The terms "supermoon" and "micromoon" are relatively new and are not strictly defined in astronomical terms. The slight variations in the moon's apparent size and brightness during these events are not always easily noticeable to the casual observer. However, supermoons, in particular, have gained popularity in recent years, as they provide an opportunity to view and photograph the moon in its most impressive and luminous state.

It's worth noting that supermoons and micromoons do not have any significant impact on Earth, beyond the slight variations in the moon's gravitational pull. While some people claim that supermoons can cause increased tidal activity or even natural disasters, there is no scientific evidence to support these assertions.

Lunar Eclipses: When the Earth Shadows the Moon

Just as solar eclipses occur when the moon passes between the sun and Earth, lunar eclipses happen when Earth passes between the sun and the moon, casting its shadow on the lunar surface. Lunar eclipses can only occur during a full moon phase, when the sun, Earth, and moon are closely aligned.

There are three types of lunar eclipses: total, partial, and penumbral. During a total lunar eclipse, Earth's umbra (the darkest part of its shadow) completely covers the moon. This can cause the moon to take on a reddish or coppery appearance, often referred to as a "blood moon." The red color is caused by Earth's atmosphere scattering and refracting sunlight, with the longer red wavelengths reaching the moon's surface.

Partial lunar eclipses occur when only a portion of the moon passes through Earth's umbra, while the rest of the moon remains in the penumbra (the lighter, outer part of Earth's shadow). The moon's appearance during a partial eclipse depends on the extent of the umbral coverage, with the eclipsed portion appearing darker or reddish.

Penumbral lunar eclipses are the least dramatic and can be difficult to observe, as they occur when the moon passes only through Earth's penumbra. During these events, the moon may appear slightly darker than usual, but the effect is much more subtle than that of a partial or total lunar eclipse.

Unlike solar eclipses, which are only visible from specific locations within the path of totality or partial visibility, lunar eclipses can be observed from anywhere on the night side of Earth, as long as the sky is clear. This makes lunar eclipses more widely accessible to people around the world.

Lunar eclipses also tend to last longer than solar eclipses, with totality often lasting for over an hour. This is because Earth's umbra is much larger than the moon's umbra, allowing the moon to remain within Earth's shadow for an extended period.

While lunar eclipses may not be as visually striking as total solar eclipses, they still offer a fascinating and accessible way to observe the intricate dance of the sun, Earth, and moon in our solar system.

Conclusion

In conclusion, the sun and moon provide us with a wealth of awe-inspiring phenomena, from the rare and dramatic total solar eclipses to the more frequent but equally captivating solar and lunar halos, supermoons, and lunar eclipses. By exploring these celestial events, we gain a deeper appreciation for the beauty and complexity of our universe.

Total solar eclipses are rare and awe-inspiring astronomical events. Their occurrences are precisely predictable, and for a given location like North America, they can be listed years in advance.

The Next 10 Total Solar Eclipses

The next 10 total solar eclipses visible from some part of North America are:

1. April 8, 2024: This highly anticipated eclipse will be visible across North America, with a path of totality

stretching from Mexico, across the United States (including Texas, Oklahoma, Arkansas, Indiana, Ohio, and New York), and into Canada (Ontario, Quebec). Millions of people are expected to witness this event.

2. August 23, 2044: Canada and the northeastern United States will have the opportunity to view this eclipse, with the path of totality crossing Montana, North Dakota, and Canadian provinces like Saskatchewan, Manitoba, Ontario, Quebec, and Newfoundland and Labrador.

3. August 12, 2045: Another significant event for the U.S., this eclipse will have a path of totality stretching from California through Florida, passing through states like Nevada, Utah, Colorado, Kansas, Missouri, Illinois, Kentucky, Tennessee, North Carolina, South Carolina, and Georgia.

4. March 30, 2052: This eclipse will be visible in parts of Asia, the Arctic, and western North America, including Alaska and northern Canada.

5. May 11, 2078: Observers in the Arctic, northeastern Canada, and Greenland will have the chance to witness this eclipse.

6. May 1, 2079: Shortly after the 2078 eclipse, this one will also be visible in parts of the Arctic, Greenland, and northeastern Canada, including Newfoundland and Labrador.

7. April 23, 2096: The northernmost regions of North America, including parts of Canada, will have visibility of this eclipse, which will also be seen in parts of Asia and the Arctic.

8. May 3, 2106: Touching parts of Asia, Europe, North Africa, Greenland, and the northeastern tip of North America, this eclipse will offer a wide range of viewing opportunities.

9. May 14, 2108: The Arctic, Greenland, and northeastern Canada will once again have the opportunity to experience totality during this eclipse.

10. March 20, 2134: This distant future eclipse will be visible in parts of Asia, Europe, North Africa, the Arctic, and the northern fringes of North America.

It is essential to keep in mind that these predictions are based on current astronomical calculations and may be subject to minor adjustments as the dates approach. Each eclipse offers a unique opportunity to witness one of nature's most spectacular phenomena, and the visibility of totality will depend on the observer's specific location within North America.

2024 Eclipse: Best Viewing in North America

Here are some of the best viewing times and locations for the April 8, 2024 total solar eclipse in different parts of North America:

1. Mazatlán, Sinaloa, Mexico
   - Start of partial eclipse: 10:13 a.m. PDT
   - Start of total eclipse: 11:07 a.m. PDT
   - Duration of totality: 4 minutes, 26 seconds
2. Austin, Texas, United States
   - Start of partial eclipse: 11:41 a.m. CDT
   - Start of total eclipse: 1:36 p.m. CDT
   - Duration of totality: 3 minutes, 48 seconds
3. Dallas, Texas, United States
   - Start of partial eclipse: 11:40 a.m. CDT
   - Start of total eclipse: 1:39 p.m. CDT
   - Duration of totality: 3 minutes, 46 seconds
4. Little Rock, Arkansas, United States
   - Start of partial eclipse: 11:47 a.m. CDT
   - Start of total eclipse: 1:50 p.m. CDT
   - Duration of totality: 3 minutes, 28 seconds
5. Indianapolis, Indiana, United States
   - Start of partial eclipse: 1:50 p.m. EDT
   - Start of total eclipse: 3:06 p.m. EDT
   - Duration of totality: 3 minutes, 48 seconds
6. Cleveland, Ohio, United States
   - Start of partial eclipse: 1:51 p.m. EDT
   - Start of total eclipse: 3:12 p.m. EDT
   - Duration of totality: 3 minutes, 46 seconds
7. Buffalo, New York, United States
   - Start of partial eclipse: 1:54 p.m. EDT

- Start of total eclipse: 3:18 p.m. EDT
- Duration of totality: 3 minutes, 36 seconds
8. Rochester, New York, United States
    - Start of partial eclipse: 1:55 p.m. EDT
    - Start of total eclipse: 3:20 p.m. EDT
    - Duration of totality: 3 minutes, 38 seconds
9. Montréal, Québec, Canada
    - Start of partial eclipse: 1:58 p.m. EDT
    - Start of total eclipse: 3:27 p.m. EDT
    - Duration of totality: 3 minutes, 22 seconds

Please note that these times are approximate and may vary slightly depending on your exact location within each city. Additionally, weather conditions can greatly impact visibility, so it's essential to monitor forecasts leading up to the event and have backup viewing plans if necessary.

2024 Eclipse: Best Viewing in Other Parts of the Globe

While the April 8, 2024 total solar eclipse will be primarily visible from North America, there are a few locations in other parts of the world where a partial solar eclipse will be observable. However, it's important to note that the eclipse will not be total in these locations.

1. Reykjavik, Iceland
    - Start of partial eclipse: 5:13 p.m. GMT
    - Maximum eclipse (36.4% coverage): 6:06 p.m. GMT
    - End of partial eclipse: 6:58 p.m. GMT
2. Lisbon, Portugal
    - Start of partial eclipse: 5:29 p.m. WEST
    - Maximum eclipse (12.1% coverage): 6:06 p.m. WEST
    - End of partial eclipse: 6:42 p.m. WEST
3. Casablanca, Morocco
    - Start of partial eclipse: 5:34 p.m. WEST

- Maximum eclipse (6.8% coverage): 6:05 p.m. WEST
- End of partial eclipse: 6:34 p.m. WEST
4. Dakar, Senegal
   - Start of partial eclipse: 4:27 p.m. GMT
   - Maximum eclipse (1.3% coverage): 4:50 p.m. GMT
   - End of partial eclipse: 5:13 p.m. GMT
5. Nuuk, Greenland
   - Start of partial eclipse: 4:27 p.m. WGT
   - Maximum eclipse (47.6% coverage): 5:28 p.m. WGT
   - End of partial eclipse: 6:26 p.m. WGT
6. Ponta Delgada, Azores, Portugal
   - Start of partial eclipse: 4:38 p.m. AZOST
   - Maximum eclipse (33.6% coverage): 5:29 p.m. AZOST
   - End of partial eclipse: 6:18 p.m. AZOST
7. Tórshavn, Faroe Islands
   - Start of partial eclipse: 5:09 p.m. WEST
   - Maximum eclipse (57.2% coverage): 6:07 p.m. WEST
   - End of partial eclipse: 7:03 p.m. WEST

Please keep in mind that these locations will not experience the dramatic effect of a total solar eclipse, as only a portion of the sun will be obscured by the moon. Nonetheless, with proper eye protection, observing a partial solar eclipse can still be an interesting astronomical event.

As we look to the future, these celestial events serve as reminders of the endless wonders that await us in the sky. By studying, observing, and appreciating these phenomena, we can foster a deeper connection with the natural world and inspire future generations to explore the mysteries of the universe. So, mark your calendars, prepare your eclipse glasses, and get ready to experience the magic and majesty of the sun and moon in the coming years.

FREE SAMPLE from this book coming up NEXT!

FREE SAMPLE (Ch.5: Exoplanets and the Search for Life)

"Light thinks it travels faster than anything but it is wrong. No matter how fast light travels, it finds the darkness has always got there first, and is waiting for it." - Terry Pratchett

Overview of Exoplanets

Exoplanets are planets that orbit stars outside our solar system. Thousands have been discovered in the last 25 years using advanced techniques, indicating planets are common in the Milky Way galaxy. Exoplanets can range from small rocky terrestrial planets like Earth to huge gas giants significantly larger than Jupiter. Many star systems have been found with multiple exoplanets orbiting their host stars. Various methods are used to detect exoplanets, including looking for transits as they pass in front of their star from our viewpoint, measuring the gravitational wobble their orbits cause in stars, gravitational microlensing effects, and direct imaging.

Exoplanets exhibit a wide range of characteristics unlike the worlds in our own solar system. Super-Earths are rocky planets bigger and more massive than Earth but smaller than ice giants like Neptune and Uranus. Hot Jupiters are gas giants orbiting extremely close to their parent star. Mini-Neptunes are medium-sized worlds surrounded by thick atmospheres of light elements. Analyzing exoplanetary properties provides clues about how planetary systems form, evolve, migrate, and interact early in their development in protoplanetary disks. The search continues for exoplanets most similar to Earth that may have conditions suitable to support alien life.

Conditions for Habitability
For an exoplanet to potentially harbor life, it must meet conditions to be habitable by having the ability to sustain liquid water on its surface over significant geological periods. The habitability of a world depends on key factors like its distance from the host star, atmospheric composition, and surface temperature. Stars like our Sun provide frequent targets in exoplanet searches focused on a habitable zone at ideal distances for life-bearing worlds.

Main sequence G dwarf stars with long stable lifetimes offer prime candidates around which habitable zone planets could evolve over billions of years. Red dwarf M stars are most abundant but can be too dim and have too long pre-main sequence phases unsuitable for life unless planets orbit very closely. For habitable surface conditions, an exoplanet should have atmospheric pressure and greenhouse gases like carbon dioxide, methane and water vapor to insulate and regulate temperatures.

Additional criteria are considered when evaluating potentially habitable exoplanet candidates, including their size, mass, density, temperatures, atmospheric loss rates, and potential for geologic activity. Best candidates are planets with active geology but not overactive volcanic activity or extreme plate tectonics that disrupt evolved complex biospheres. Life likely requires renewed sources of chemical energy over time and protection from harsh space and stellar radiation.

Quotes About Alien Life

The concept of alien life stirs the imagination and provokes a multitude of questions about our place in the universe. The following quotes reflect the intrigue and curiosity that the possibility of extraterrestrial beings inspires in scientists, writers, and thinkers alike:

- "The universe is a pretty big place. If it's just us, seems like an awful waste of space." - Carl Sagan
- "If there is one thing the history of evolution on Earth has taught us it's that life will not be contained." - Michael Crichton
- "Perhaps we've never been visited by aliens because they have looked upon Earth and decided there's no sign of intelligent life." - Neil deGrasse Tyson
- "It's not a matter of if, but when we will find alien life." - Michio Kaku
- "The discovery of extraterrestrial life, perhaps more than any other single event, would alter our concept of who we are and where we stand in the cosmic scheme of things." - James Trefil
- "Not to search for life elsewhere would be a failure of human curiosity." - Reza Aslan
- "It is possible that the future of human civilization depends on the ethical treatment of extraterrestrial life forms." - Nick Bostrom
- "The high probability of the existence of life outside our planet makes the search for extraterrestrial intelligence one that we must pursue." - Carol Stoker
- "If we ever encounter extraterrestrial intelligence, I believe it is overwhelmingly likely to be post-biological in nature." - Paul Davies
- "Aliens might be staring at us in disbelief that we cannot see them." - Seth Shostak
- "The thought of being the only intelligent species in a dumb universe is the second worst existential nightmare.

The worst is being watched by an intelligence greater than our own." - Charles Stross
- "Life is a miracle, and it doesn't have to be confined to Earth-like conditions to exist." - Avi Loeb

These quotes express a range of perspectives on the search for and potential impact of discovering life beyond Earth. They reflect a mix of optimism, caution, humor, and profound contemplation about what such a discovery would mean for humanity's understanding of life itself and our place in the cosmos.

Recent Exoplanet Discoveries
- NASA's Kepler space telescope mission identified over 2,600 confirmed exoplanets along with thousands of additional candidates from observing segments of the night sky.
- The Transiting Exoplanet Survey Satellite (TESS) has found hundreds of exoplanets transiting nearby bright stars, providing prime targets for further detailed study and characterization.
- Ground-based searches using precise radial velocity measurements continue to unveil exoplanets by the wobble they induce in their star's motion from gravitational tugs.
- New space telescopes like CHEOPS and the upcoming James Webb are designed to precisely analyze exoplanet atmospheres for molecular signatures and conditions indicative of habitability.
- Improved observations indicate approximately 1 in 5 Sun-like stars have an Earth-sized planet orbiting in their habitable zones at distances for liquid surface water.
- Some recent potentially habitable discoveries include the exoplanets K2-18b, pi Mensae c, TOI-700d and LHS 1140 b, all orbiting M dwarf stars and possibly having liquid water under their surfaces.

Ongoing SETI Research
The search for extraterrestrial intelligence (SETI) involves experiments attempting to detect signs of advanced alien

civilizations through interstellar radio signals, optical flashes, probes, spacecraft, and other means.

- Optical and radio telescopes like the Allen Telescope Array and facilities used by the Breakthrough Listen initiative monitor millions of stars and scan large swaths of the sky seeking modulated signals or energy flashes that could indicate technology.
- New directed messaging efforts like METI (Messaging Extraterrestrial Intelligence) have transmitted messages and images like the Arecibo Message via radio telescopes out towards promising star systems to attempt establishing contact.
- Advances in data analysis and artificial intelligence use machine learning algorithms to search enormous sets of radio telescope data seeking anomalous patterns that could represent potential alien communications.
- No definitive evidence of extraterrestrial civilizations through signals or communication has been found in searches so far. But SETI research remains ongoing and holds immense discovery potential if contact is ever made in the vast cosmos.
- Some scientists caution that replying to unknown signals or advertising human presence widely could be risky before better understanding the motivation and nature of alien intelligences.

The possibility of life beyond Earth captivates human imagination and scientific curiosity alike. Exoplanet statistics indicate billions of potentially habitable worlds likely exist in the Milky Way galaxy alone. Future telescopes will thoroughly probe their atmospheres and surfaces. We may stand on the verge of uncovering signs of life beyond our solar system.

A Final Word

I truly appreciate your participation in this unique journey through the mysteries of physics. If you liked this book, please help me spread the word by:

- Leaving a 5-star review at the store where you purchased this book.
- Telling your siblings, classmates, friends and relatives about this book
- Recommending this book to your teacher, coach or educator, and
- Sharing your thoughts on social media

I hope you also liked the free sample from the book, *Astronomy Nerd*. Do check out our other exciting titles and stay tuned for new and exciting releases from Lexicon Labs. Some of them are highlighted in the pages that follow.

I wish you lots of good luck and new adventures!

Dr. Leo Lexicon

Advice you always wanted, but were too afraid to ask!

SCAN ME

Discover the 10 Life Hacks that will:

- Unlock the Secrets to Teen Empowerment
- Boost Motivation & Achievement
- Deepen Parent-Teen Connections
- Shape a Bright, Successful Future
- Build a life of purpose and fulfillment

Follow Dr. Leo Lexicon on Twitter/X

 X @LeoLexicon

LEXICON LABS

For the Science Nerds
OTHER TITLES FROM THIS SERIES

- Entertaining explanations of core concepts
- Mind-blowing concepts and applications of physics and chemistry that you never knew were possible
- Survey of the latest breakthroughs
- Amazing facts and quotes from the top minds in the field
- Perfect travel companion or gift

Follow Dr. Leo Lexicon on Twitter/X

 X @LeoLexicon

AI FOR YOUNGER READERS
Get on the leading edge of the AI revolution!

- Perfect for readers Ages 6-9
- Structured introduction to the building blocks of AI
- AI concepts explained in a simple, easy-to-understand format by a Bay Area educator
- Resources for puzzles, games, and coding
- Perfect travel companion or gift

Follow Dr. Leo Lexicon on Twitter/X

 𝕏 **@LeoLexicon**

Discover More Bestselling Titles from Lexicon Labs!

SCAN ME